Planetary Defense:
On Near Earth Objects
Threat Mitigation

I. Introduction

The issue of threats of impacts from Near Earth Objects and other celestial bodies is a topic which has been addressed many times before and will very likely continue to be so. The present paper was motivated by the release of the draft Final Report on Near Earth Object (NEO) study, prepared by NASA for the U.S. Congress in December 2006; this document has been labeled by NASA as "pre-decisional" and only a draft – a status that has not yet changed by this date (Oct. 2007). In 2005, the U.S. Congress directed NASA to perform an analysis of alternatives to "detect, [...] characterize potentially hazardous objects (PHO), and submit an analysis of alternatives for threat mitigation". That 275-page document contains the summary of several years of studies performed by multiple investigators and published or presented elsewhere. As such, the NASA report is an excellent summary introduction to the overall challenge presented by NEO threats, feasible technical approaches, and logistical and budgetary consequences. The present memorandum focuses on the potential role of USAF and the relationship with advanced technology development programs that are sponsored by the USAF or are candidate for such sponsorship. Related recommendations are also offered in the final section. First a summary of the NASA report is provided, followed by a discussion of technical approaches, and a brief analysis of future technology requirements. While there can be many geo-political implications of the threat itself and approaches to its remediation, the present paper does not attempt to discuss these in great detail; they will be mentioned only as contextual elements of the technical discussion. It is also important to emphasize that this document does not represent a complete analysis of the technical requirements, nor does it describe an official position by the USAF; there are currently too many unknowns preventing such decisive statements and important decisions involving Planetary Defense can only be made at a high level of authority.

II. Threat Definition

The threat posed by the impact of asteroids and comets with the Earth has been well publicized, with varying degrees of accuracy, by the entertainment industry. A more definite assessment of the threat can be made from studies of prior impacts, starting from a major event 65 millions years ago – the Cretaceous/tertiary (K/T) event – in what is now the tip of the Yucatan peninsula (Chicxulub crater) and down to smaller, presumed events that correlate with historical records of climate change. There is now a majority opinion that the Yucatan event is at least a major contributor, if not the direct cause of the extinction of the dinosaurs. Moreover, it is clear that NEO impacts can cause great havoc on human activity, from economic disruptions to significant loss of life, climate change, "end-of-civilization", or complete extinction of the human race. The probability of these events decreases with the severity of the impact, and size (mass) of the NEO. Figure 1 and Table 1, shown below and taken from [1], illustrate the range of event probability and impact consequences as function of the NEO size.

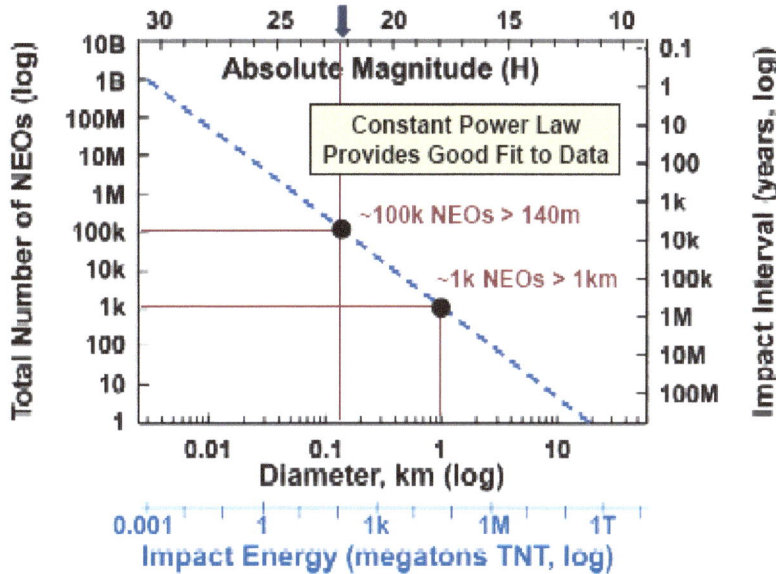

Figure 1: Frequency of NEOs by size, impact energy and magnitude (taken from [1]).

Type of Event	Diameter of Object	Fatalities per Impact	Typical Impact Interval (years)
High altitude break-up	< 50 m	~0	annual
Tunguska-like event	> 50 m	~5,000	250 - 500
Regional event	> 140 m	~50,000	5,000
Large sub-global event	> 300 m	~500,000	25,000
Low global effect	> 600 m	> 5 M	70,000
Nominal global effect	> 1 km	> 1 B	1 million
High global effect	> 5 km	> 2 B	6 million
Extinction-class Event	> 10 km	6 B	100 million

Table 1: Impact Frequencies and Typical Consequences (taken from [1]).

We add three important observations:

(1) The total number of NEOs has by itself been a variable; from 1800 to modern times, the number of NEOs discovered has risen dramatically as a result of more capable telescopes and more observations. This tends to shift the curve of Figure 1 upwards; if the size distribution is not affected, one would then conclude that more Potentially Hazardous Objects – PHO (i.e. NEOs above a critical size) would be discovered, and the probability of impact would be increased. However, there is a limitation to this argument, since large NEOs are easier to detect; thus, it is more reasonable to infer that all the large NEOs can be catalogued within a reasonable time, while smaller and less consequential objects are added to the list as both detection technology and observation times improve.

(2) The statistics of rare events obey the Poisson distribution

$$f(k;\lambda) = \frac{e^{-\lambda} \lambda^k}{k!} \qquad (1)$$

where k is the number of events and λ is the expected number for a given time of observation. For example according to Table 1 above, one should expect one impact by an object larger than 1 km every 1 million years (1 Myr); the expected number of such events within the next 20 yrs is therefore: $\lambda = 20\,\text{yrs}\,/\,1\,\text{Myr} = 2 \cdot 10^{-5}$. One can then construct another table that estimates the *average* number of fatalities[1] within the next 20 yrs and 100 yrs:

Ø	Fatalities (20 yrs)	Fatalities (100 yrs)
< 50 m	0	0
> 50 m	750	3400
> 140 m	540	2700
> 300 m	1,100	5400
> 600 m	3,900	19,400
> 1 km	54,000	272,000
> 5 km	18,000	90,000
> 10 km	3,300	16,000

Table 2: Expected fatalities from NEO impacts within 20 and 100 yrs periods.

One can point out that these numbers are extremely low, about an order of magnitude lower than fatalities from automobile accidents nationwide; therefore, on the basis of *average* numbers of fatalities expected from NEO impacts, there seems to be little reason for the U.S. Congress to be overly concerned and spend significant resources on this issue. However, the problem is not with the *average* number of fatalities, but the *peak* number and associated consequences at each event. While the country could certainly recover from impacts of NEOs of diameter Ø=300 m and below, even up to 600 m diameter given sufficient time[2], there is no recovery possible from impacts by larger size NEOs, and the day an "extinction"-class event occurs, there is no safety in deep underground bunkers. The last event of this category was 65 Myr ago, and the probability of having had no similar impact in that period of time is approximately 50%. Humans (arguably all mammals in general) can therefore consider themselves reasonably fortunate not to have been wiped out of existence during that gap.

(3) There are still significant uncertainties in the estimated number of PHOs and probability of collisions. There are also IEOs (Interior Earth objects, whose orbit is *almost* entirely within the Earth orbit) and so-called Aten asteroids, which spend most of their time within the Earth orbit but are crossing it. The rate of active comets dropping within the inner solar system is estimated from historical records to be very low, but given the poor statistics and complete lack of knowledge of the population of such objects in the Kuiper belt, is very much uncertain[3]. Some authors also claim that the impact rate is significantly under-estimated and should be revised upwards [2].

---

[1] These numbers assume a constant population; using the most recent growth trend, the adjusted numbers should be 40% higher for the 20 years case, while for 100 yrs the numbers would be more than 4 times larger, in which case the world population would have far exceeded a sustainable level.

[2] By comparison, the number of fatalities during WWII was approximately 72 million worldwide, 400,000 for the US and 23 million for the Soviet Union. While the casualties from a 600m NEO impact are less than 2% of the US population, the catastrophic effect on the economy would likely lead to societal failure.

[3] Note that all comet impacts have at least a high global effect if not extinction-class events, due to their significant mass but also their much higher kinetic energy compared to asteroids. Furthermore, comets are more likely to break-up from the Sun's gravitational influence – in which case, even if the main body

There is clearly a threshold level (in size, impact energy, estimated casualties or material damage) below which it no longer makes sense to be concerned about threat remediation. A detailed cost-benefit analysis could determine this cut-off level, although with considerable uncertainty. Ultimately, this decision must be made by the U.S. Congress or some international organization if the threat-remediation campaign is part of a larger community. For the NASA study, no specific threshold was provided by Congress, and the study group chose an object diameter of 140 m, a reasonable number.

At this stage, we can tentatively divide the threat from NEOs into the following classes:
(1) Small objects: these are inconsequential (i.e. they mostly burn-up in the atmosphere and produce very limited damage), or fall below the threshold of a cost-benefit analysis, yet TBD.
(2) Medium objects: these are above the threshold for concern, but do not constitute a global threat, i.e. which do not endanger modern civilization or the human race.
(3) Large objects: these pose a major threat to survival of civilization or species.

We could also point out a 4th class of threats at the extreme range of size, i.e. very large objects for which there are no known counter-measures (they are of course well above the extinction-class). These "planet-killers" (such as the collision event which would have created the Moon) fortunately disappeared from the solar system or are well-known[4]; being no-longer a threat, they are relegated to the realm of fiction (e.g. [4]).

The diameter threshold between the first and second classes can be set to approximately 140 m, following [1], while the threshold between second and third categories is more of the order of 1 km. It is the class-3 ("Large") impact which is of concern here. One should not be lulled into a false sense of security by the extreme dilution of the probability distribution of very rare events[5] – given the death statistics, any person living in the US can expect to live an average of 50,000 years before being killed by a lightning strike, yet most people take reasonable precautions against such an event. While impacts from smaller objects are more probable within the next few decades, we can always survive those if caught unprepared. It has been repeatedly suggested elsewhere that the effort in NEO threat mitigation should be focused on small objects (< 100 m) (e.g. [5]); while the much higher frequency of such impacts makes the threat more understandable to the general public, the argument that this would result in an active program is highly doubtful. The relatively minor consequences of such an impact inevitably force a comparison with other natural (hurricanes, earthquakes, tsunamis, volcanic eruptions) or man-made (acts of terrorism) event categories, with a lobbying constituency behind each. Given limited resources, it would not be reasonable to expect the U.S. Congress to be more amenable to fund a program preventing an impact by a 100 m asteroid – likely to land in some unpopulated area – as opposed to hurricane or earthquake preparation. On the other hand, there is no comparison to the sheer magnitude of an extinction-class

---

misses the Earth, accompanying debris may not – or massive outgassing, which also makes the comet trajectory highly unpredictable.

[4] The Asteroid Ceres, for example, is the largest known at about 1000 km in diameter.

[5] The likelihood of an extinction event in a given year may be very small, but given sufficient time this will eventually occur. Given the Poisson probability distribution, the odds to have two extinction events within 65 Myr (time at which the last one occurred), given one impact per 100 Myr in average (if accurate), is about 11%, and the odds of that event are essentially the same for 2008 than 10,000 yrs from now.

impact. The very low probability of such an event makes it equally difficult for leaders, even with extraordinary vision, to support a mitigation program against this class of threats; nevertheless, it would be prudent to do so, and focusing on minor threats may not enable us to prevent the disaster that really counts…

Ideally we should have a strategy to design counter-measures applicable to major threats as soon as possible, while developing approaches for dealing with small objects in a more cost-effective manner, i.e. moving the threshold between classes 1 and 2 towards the smaller objects. This helps defining the overall strategy for threat remediation:

(a) Large objects- Since the damage caused by such impacts is catastrophic, cost should not be an issue when suddenly faced with an identified threat. Instead, *effectiveness* of the remediation approach and the *rapidity* of the response are critical. Such threats could be from large NEOs, which are likely to be discovered well in advance if a solid detection campaign was put in place, but the threat could also be a comet-type object, for which there can be insufficient warning.

(b) Medium objects- The damage of such impacts is survivable, while the probability of damaging events is much more substantial than in the previous case. Inevitably, the cost/benefit ratio is the primary consideration, however callous it may sound. While a cataloguing and mitigation strategy is being developed and costs become lower, the threshold of the lower-size of the spectrum can decrease.

Thus, threat remediation implies both a short-term and long-term strategies, which will be discussed in further detail below. As principal objectives of this paper, we can also identify the key role that can potentially be played by DOD in this mission.

III. Technical Solutions

III-A. Detection and Tracking

As identified in the NASA report [1], NEOs must be detected, tracked and cataloged. Detection is currently made mostly by visual means, using the sunlight reflected from the object as it moves against a fixed background of stars. The ease of detection depends on the object size, its average albedo, and distance from the Sun. Once detected and identified as an asteroid, it must be tracked to determine its trajectory with as much precision as *necessary*. Tracking accuracy depends on the observation time; since many NEOs and PHOs need to be identified, it would be counter-productive to spend time and resources increasing the accuracy of orbits if the objects pose no threat for a foreseeable future. The uncertainty of the trajectory can be shown as ellipses centered on the object position, each ellipse being a contour for a given risk; the trajectory becomes then a "tube", as shown in Figure 2 with growing thickness as time progresses. If the acceptable risk of impact is (somewhat arbitrarily) set at 1/1 million, for example, the trajectory must be defined with sufficient accuracy that the Earth lays outside of the surface of the tube for that level of orbit accuracy[6].

This should not be confused with the *apparent* low-risk associated with extreme dilution of the probability density function when the uncertainty is large. Consider for example the initial discovery of a NEO; the time spent tracking the object is minimal, and the error ellipse can be extremely large, of the order of the size of the inner solar system. The

---

[6] For a normal distribution and a 1/1 million risk, this means that the Earth is outside a 5-sigma envelope.

trajectory can take any path within this space and the probability that it intersects that of the Earth is very low. However, this is only an artifact of the lack of knowledge about the object. As further measurements are taken, the error ellipse rapidly decreases in size and if the Earth still remains within it, the risk of collision increases correspondingly, until sufficient precision is obtained that the Earth is found to be outside the critical envelope. Over time, the risk of collision by a given NEO would therefore increase gradually until it suddenly drops to almost zero. This has led many to question the wisdom of notifying the population of all such events, since many "false alarms" may be raised unnecessarily. Note also that the error ellipse also extends *along* the trajectory line (not shown in Figure 2), since even if the orbits of the NEO and Earth intersect, an impact requires that both objects be co-located *at the same time*.

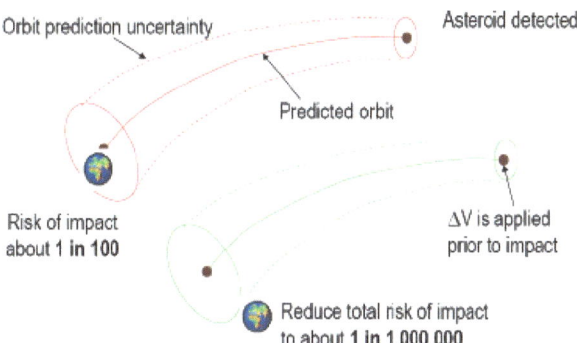

Figure 2: PHO approach uncertainties and deflection goal (taken from [1]).

The NEOs can be detected through optical means using visual or infrared (IR) spectral ranges using wide-area sky surveys, and using ground or space-based telescopes. An extensive summary of the respective benefits of various approaches to detection and cataloguing of NEOs is reported in [1]. Suffice it to say that an optimal approach yielding the fastest rate of NEO identification, including the IEOs and Aten asteroids mentioned earlier consists of telescopes, preferably IR, in an inner sun-centric orbit such as the Venus orbit or the inner Sun-Earth Lagrange point L1.

Although accurate tracking of the NEOs can be greatly facilitated by radar, its signal typically decays as the fourth power of the distance ($1/r^2$ for emitted signal, and another $1/r^2$ factor for reflected signal), and this is obviously not a practical solution except at relatively close range. This solution was reserved in [1] as an additional mean of obtaining very accurate trajectory data for a few objects of interest. However, the report appears to ignore the potential of phased-array antennas with highly directional beams or masers; this would essentially eliminate the outward $1/r^2$ signal decay and considerably extend the range for accurate tracking or radar imaging. High accuracy of detection over long periods of time for objects of potential interest could also be obtained by intercepting the NEO and anchoring a transponder on its surface. This is especially attractive for NEOs that are tied to a "keyhole", such as the projected encounter of the asteroid Apophis in the vicinity of earth in 2029. Such NEO rendezvous missions can take several years, even with the most advanced power and propulsion technology on the design boards. An extensive campaign to analyze and characterize the NEOs in-situ would be a long-duration endeavor; if such a detailed characterization is a prerequisite to the mitigation, less time is available to successfully perform the deflection. On the other hand, any remote (optical, IR, radar) means of characterization would increase the time available for the deflection.

Thus, the placement of space platforms in appropriate orbits, as well as routine missions to PHOs could greatly accelerate and improve the detection rate and enhance the tracking accuracy of NEOs. This brings both challenges and opportunities, since the space observatories must be deployed and be operational for a sufficient time, be highly automated and highly reliable (it is difficult to envision manned missions to L1 to service an IR telescope). Robotic missions to such space observatories could be highly beneficial if they can be achieved reasonably cheaply. "Routine" interception of NEOs for object tracking (transponder) and characterization also implies the ability to send multiple spacecrafts on high delta-V missions with very low cost. Space-based phased-array radar antennas require the ability to deploy and assemble large structures, again at low cost. Taken together, these requirements imply the need for an efficient infrastructure for deep space operations, with projected technology requirements in power, propulsion, reliability, and robotic (AI) operations.

III-B. Characterization

After detection and tracking, the second stage of a deflection campaign concerns the cataloguing and characterization of the object. This phase may be necessary for some of the technical approaches considered, as shown in Tables 3a and 3b. Mass is of course always a needed parameter, which is obtained from the analysis of the trajectory. High-accuracy optical and IR imaging can remotely determine size and shape for the larger objects, albedo can be estimated and bounds on material composition and characteristics could possibly be inferred from other data (e.g. spinning rates), albeit with large uncertainty. Detailed characterization is possible only with missions to the NEO, at least flybys, preferably involving an impactor, and ideally a landing on the NEO and surface analysis and core drilling. As implied earlier, these are high ΔV missions, and they must be performed at very low cost in order to visit as many NEOs as necessary. Power requirements for the diagnostic equipment should be relatively small and could be handled with a small or a single RTG (Radioisotope Thermal Generator). It may be preferable to have multiple components of the spacecraft, i.e. a small impactor, a small lander with a transponder, and a pair of orbital sensors for stereoscopic imaging. The trend would therefore be to design specialized pico-satellites that can be released at the NEO approach and perform their individual functions.

One could consider several approaches to a systematic campaign of NEO characterization: first, each NEO rendezvous could be a separate mission and the propulsion system to provide the ΔV required could be conventional and jettisoned (Hohmann transfer); second, each mission is provided with a highly efficient, low-power EP system and the intercepting trajectory is a low-thrust spiral; third, several NEOs are visited by a single spacecraft/space-tug that delivers the monitoring pico-sats for each rendezvous, and that uses an advanced, high-power propulsion system. The respective benefits of each approach depend on a number of factors, including the overall long-term strategy involving NEO cataloguing and deflection campaign.

	Mass	Spin	Density	Material Properties	Size & Shape	Surface Properties
Conventional Expl. Surface - Contact	Yes	No	Helpful	Helpful	Helpful	Helpful
Conventional Expl. Subsurface	Yes	No	Helpful	Helpful	No	No
Kinetic Impactor	Yes	No	Helpful	Helpful	Helpful	No
Nuclear (Contact)	Yes	No	Helpful	Helpful	Helpful	No
Nuclear (Standoff)	Yes	No	No	No	No	No
Nuclear Explosive (Sub-Surface)	Yes	No	Helpful	Helpful	No	No
Nuclear Explosive (Surface Delayed)	Yes	Yes	Helpful	Helpful	No	Helpful

Table 3a: Characterization required for impulsive alternatives (taken from [1]).

	Mass	Spin	Density	Material Properties	Size & Shape	Surface Properties
Yarkovsky	Yes	Yes	No	No	Yes	Yes
Focused Solar	Yes	Helpful	No	No	No	Yes
Gravity Tractor	Yes	Yes	No	No	Yes	No
Mass Driver	Yes	Yes	Yes	Yes	Helpful	Helpful
Pulsed Laser	Yes	Helpful	No	No	No	Yes
Space Tug	Yes	Yes	No	No	Yes	Yes

Table 3b: Characterization required for slow push alternatives.

III-C. Deflection
III.C-1. Nuclear & Impulsive Approaches

It is worth pointing out from Table 3 that the only deflection approach that requires no further characterization of the NEO besides its mass (determined from optical observation of its trajectory), is the nuclear stand-off. In this case, the weapon is detonated at some distance from the NEO, and the radiative energy of the detonation (mostly X-rays and neutrons) is used to ablate material on the surface of the asteroid, with the subsequent recoil providing the impulse. Because the energy is delivered as radiation, the impulse can be distributed over a large area (in contrast to high velocity impactors) and to NEOs of any shape (including rubble piles and sand piles). Furthermore, if the weapon is at sufficient distance, impulse could be provided to NEO satellites, i.e. composite objects. Finally, X-rays are absorbed within a thin layer of material and the absorption has little dependence on material surface properties. Therefore, the nuclear stand-off approach is ideal when there is insufficient time for characterization, and when remaining debris is of significant concern (i.e. little time left before impact). There are, however, some additional complications. In the case of a standard nuclear weapon, most (70%) of the energy is in the form of X-rays [6], which would heat a small amount of material (surface layer) to very high temperatures, creating an expanding plasma layer. The disadvantage of this approach is that a lot of energy can be re-radiated – a net loss – and the impulse-to-energy ratio can be relatively small –

scaling approximately as $1/T^{1/2}$. Therefore, heating a small mass to very high temperatures is not an efficient way to provide the necessary impulse to the asteroid[7]. The fact that a nuclear stand-off approach is still an attractive option is mostly the result of the enormous amount of energy available in thermo-nuclear devices. Since we are not concerned with having a limited amount of mass available, the best use of the available energy would be to heat a greater amount of material, but to temperatures just high enough to vaporize it. However, the energy threshold required depends on material properties (composition, structure, porosity, etc.), and this would probably require a prior characterization mission.

Gennery and Holsapple [7,8] suggested that neutrons would be more effective than X-rays, because the absorption length is greater; more mass would be absorbing the energy and the efficiency would be greater. Again, the specific energy must be above a material-specific threshold, and if the absorption length is too large, the effectiveness drops rapidly. Calculations by Holsapple for SiO_2 suggest that material porosity ceases to be a factor when the specific energy is approximately 100 MJ/kg; this is well above the threshold for vaporization, and in that regime the impulse-to-energy ratio scales as the inverse square-root of the temperature (or specific energy), as in the case of X-ray irradiation. Thus, peak values of impulse efficiency cannot be obtained with neutron irradiation unless better characterization of the material (preliminary rendezvous mission) is obtained. From [7], one can estimate the required yield of an optimized nuclear device for a given threat. Consider an extinction-class object with a diameter of 1 km; using the average range of 150 kg/m^2 for neutrons given in [1], average geometric irradiation factors and a stand-off distance of half the object's diameter, one needs a neutron yield of $5 \cdot 10^7$ J in order to achieve the material-independent value of specific energy deposition of 100 MJ/kg. Since in an optimized device the neutron yield is approximately 10% of the total energy, this is equivalent to a weapon yield of 1,200 megatons (Mt-TNT), or 130 times the yield of the most powerful weapon developed by the US[8]. The weapon itself would have a mass of approximately 350 metric tons[9]. Exploding the device at closer range (nearer the surface) does not necessarily improve the situation, since the maximum impulse for very high yields is obtained for a stand-off distance of approximately 1/3 the diameter, instead of ½ considered above. Furthermore, the object will have an irregular shape and it may be difficult to control very precisely the stand-off distance. Since the X-ray yield is typically larger by at least one order of magnitude, it is debatable whether it is worth attempting to optimize the coupling efficiency by relying on neutron irradiation.

Buried detonation would be a more effective approach, since the entire energy of the device can be absorbed by the target (a stand-off detonation would waste at least half of the energy in radiation that is not intercepted by the target). Re-radiation of high-temperature plasma is no longer a net-loss, since it would also be absorbed by the surrounding material. The effect most easily obtained, i.e. requiring the least energy,

---

[7] The same observations apply to the case of laser ablation.
[8] The largest nuclear device developed and tested by the Soviet Union, the "Tsar Bomba", had a yield of 50 Mt-TNT. This extraordinary requirement is also consistent with the analysis of Gennery, who considered a threat scenario necessitating an impulse of 4×10^9 kg m/s, which can be mitigated with a yield of about 10 Mt. Increasing the impulse requirement by 2 orders of magnitude (scenario F in Figure 4) yields 1,000 Mt.
[9] Assuming the same scaling relation as for existing designs.

would be fragmentation of the object; however, this is not sufficient since at the threshold of fragmentation, all the debris would remain on essentially the same trajectory. For fragmentation to be effective, the debris must have a significant transverse velocity as a result of the detonation, so that by the time of impact they are widely scattered and the probability of actual impact is greatly reduced. The minimum energy for fragmentation of a 1 km-diameter asteroid was approximated by Ahrens et al. [9] as 1 Mt-TNT; this figure is easily increased by one order of magnitude to provide the required scatter velocity to the debris (depending on the time-to-impact). This still makes a deeply buried nuclear detonation about two orders of magnitude more efficient than stand-off detonation and brings the required yield down to values where stockpiled weapons are useful. However, the problem is that a buried device will require: (a) a rendezvous mission instead of simply intercepting the NEO; (b) drilling into the core to significant (100s m) depth, a very difficult prospect for nickel-iron objects, for example, or (c) a deep impactor with delayed fuse, possible only for very soft targets. Therefore, although a buried nuclear detonation would not require development of new, very high-yield warheads, the mission time would be significantly longer and would still be extremely difficult for some targets.

Deflection by kinetic impact is also possible for relatively small asteroids. Holsapple [10] examined the performance of kinetic impactors, but based is analysis on kinetic energy considerations; this is incorrect, since kinetic energy is not conserved in inelastic collisions. For such cases, conservation of momentum is used to evaluate the impulse transferred to the main asteroid body, i.e. $\Delta I = M \Delta V = m_p v_p$. Considering for example the scenario "D" of Figure 4 for which the critical impulse is of the order of 5×10^8 kg.m/s, we find that the projectile mass must be 50 tons if the final impact velocity is 10 km/s. The constraints on the vehicle can be expressed as follows:

$$M_0 = \frac{\Delta I}{v_p} e^{v_p/v_e} < 100 \text{ tons} \qquad (2a)$$

and
$$\Delta t = \left(\frac{v_e^2/2}{\eta P_e}\right) M_0 \left(1 - e^{-v_p/v_e}\right) < 1 \text{ year} \qquad (2b)$$

where M_0 is the initial mass in orbit and $v_e = g I_{sp}$ is the exhaust velocity of the propulsion system. One can always reduce the vehicle mass with increased specific impulse, but the power requirement may become significant. In the case of $v_p = 10 \text{ km/s}$ again, the mass in orbit and minimum power requirements (for $\Delta t \approx 1$ yr) in Figure 6 show a range of specific impulse for which the initial mass is below 100 tons, and the power required is less than 1 MWe. Conventional thrusters ($I_{sp} \approx 2$ ksec) can be used but even in that case, the power requirement is at least of the order of 400 kWe; since the final (dry) mass is of the order of 60 tons, the system-α must be less than 150 kg/kWe, an easy requirement with state-of-the-art technology, and which can be addressed by nuclear reactors or very large solar arrays. Chemical propulsion results in excessively large vehicle masses; the minimum specific impulse satisfying the requirement (2a) for a final impact velocity of 10 km/s is approximately 1,350 sec, i.e. well above chemical or even nuclear thermal propulsion. Decreasing the final impact velocity for the given impulse requirement raises the specific impulse further (for this particular example, no solution to (2a) is found below $v_p \leq 5 \text{ km/s}$.

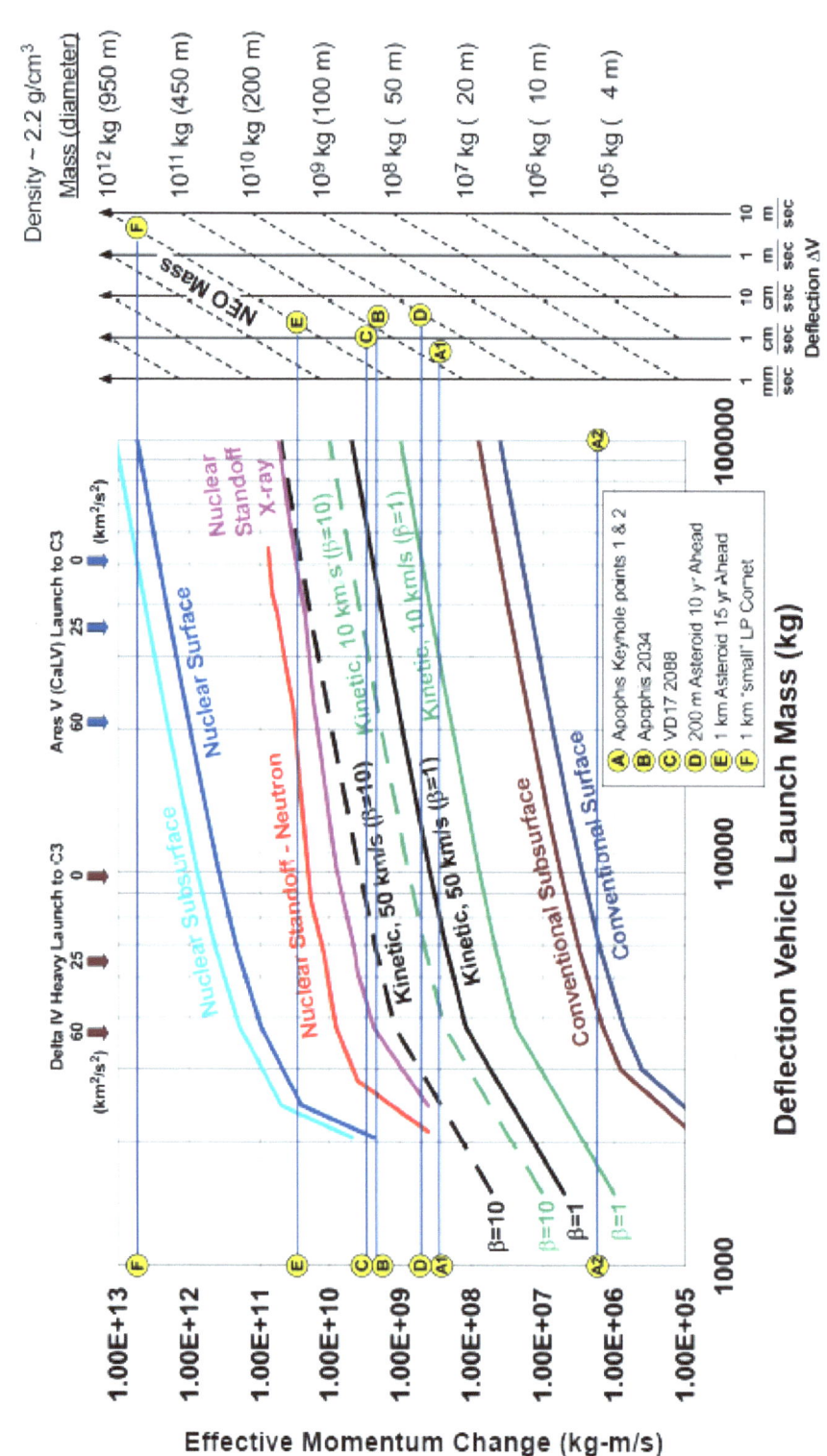

Figure 4: Momentum capability of impulsive methods for various scenarios (taken from [1]).

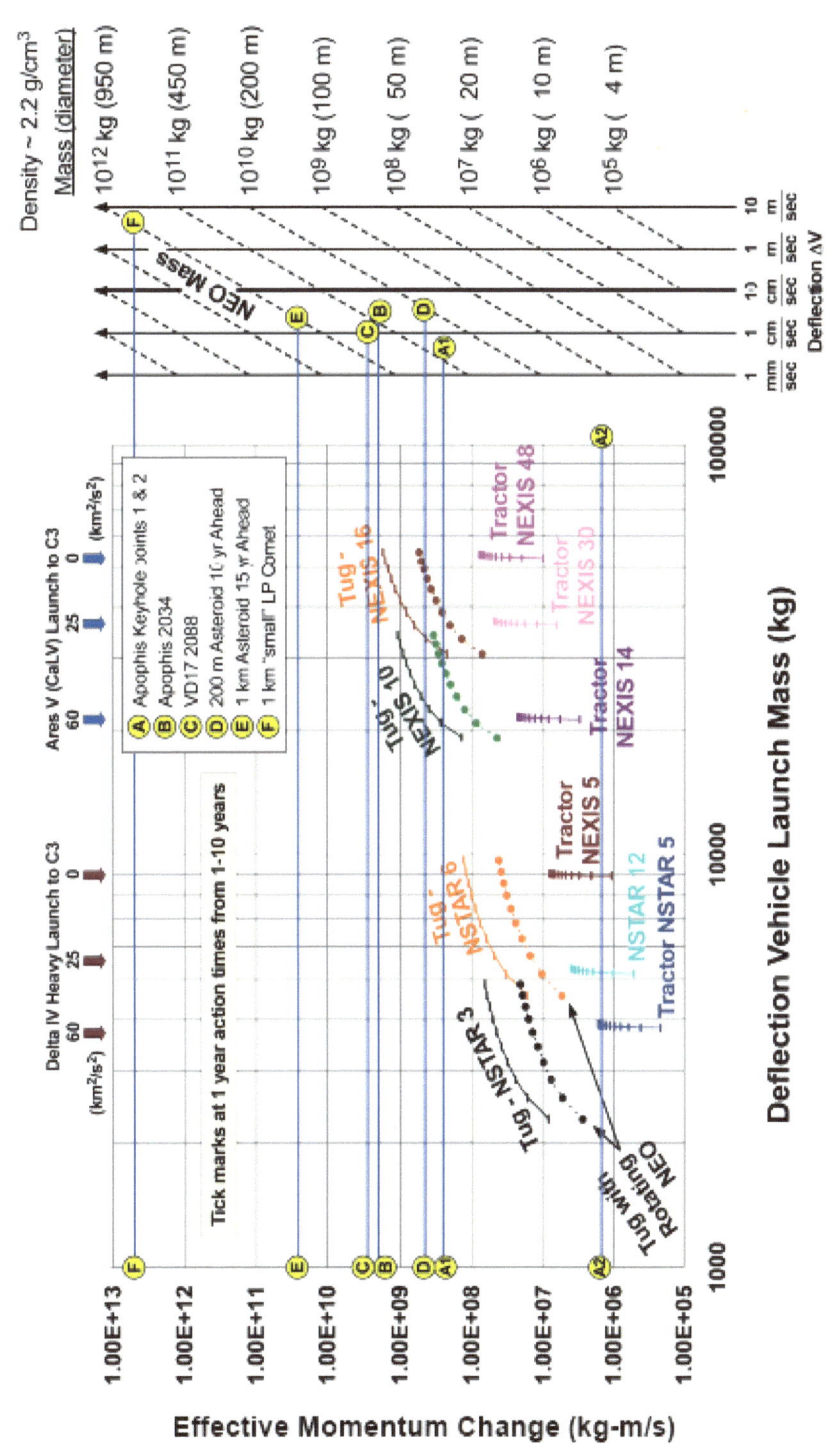

Figure 5: Momentum capability of slow-push methods for various scenarios (taken from [1]).

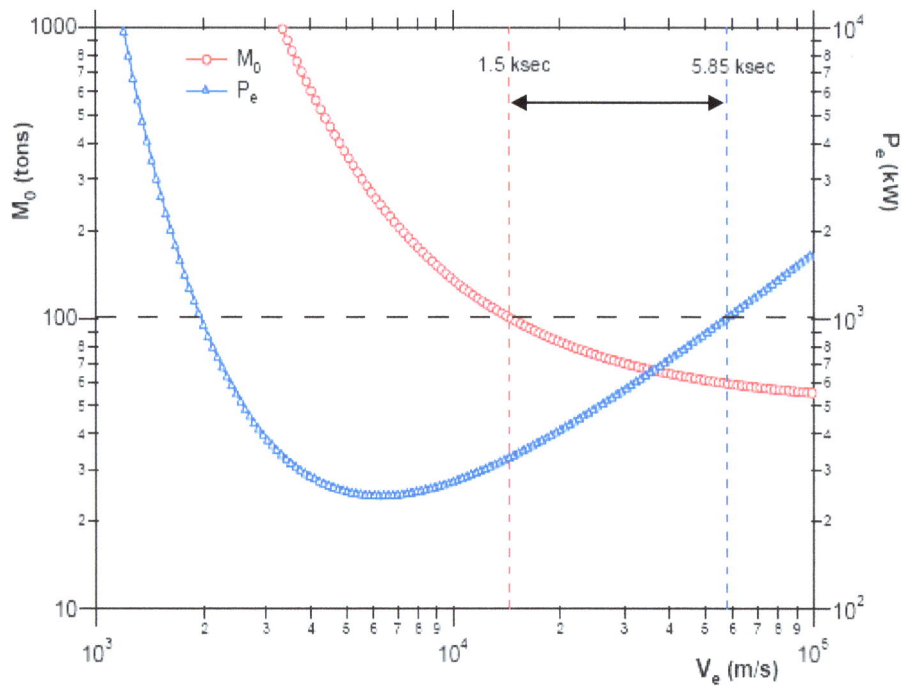

Figure 6: Example of constraints on vehicle system design for kinetic interceptor.

III.C-2. Slow-Push Approaches

Since even nuclear weapons may have great difficulties in achieving the difficult missions, it is hardly worth considering chemical explosives or kinetic energy impacts from hyper-velocity interceptors; nothing matches the energy density of nuclear weapons. The other alternatives, therefore, consist of "slow-push" methods, potentially more efficient but, as the name implies, requiring longer times. The respective performances of various trajectory deflection techniques are shown in Figures 4 and 5, for both impulsive and slow-push methods. To understand these figures, one must follow the curves for each technique until they reach above the horizontal line corresponding to the required impulse that must be given to the corresponding threat. For example, for scenario "D" (200 m asteroid with 10 years available for performing the deflection maneuver), an impulse of approximately $5 \cdot 10^8$ kg.m/s is required. The horizontal axis is the total mass available for the deflection vehicle, i.e. the mass that is launched into the intercept trajectory. If interception requires a high-energy trajectory (C3 increases), more mass is required as propellant to boost the intercepting vehicle on the intercept trajectory, leaving less mass available for the vehicle itself to complete the deflection[10]. Of course, these figures depend critically on the assumptions made about the system performance of the launcher and interceptor vehicles, and need to be revised accordingly as better technology is being developed, or if other systems (e.g. existing high-thrust OTV) are being leveraged.

Some key observations are as follows:

---

[10] It could also be pointed out that after deflection, the interceptor must also be able to avoid an impact with the Earth as well; since in a slow-push approach the vehicle is connected to, or in the immediate vicinity of the NEO, this is not really a problem.

– The gravity tractor fails to meet all requirements except for the easiest mission (deflection by 5 km only). In fact, it is likely that the gravity tractor's actual performance is even lower than assumed, due to the need for dynamic stability control and increased propellant requirements.

– The space tug can meet the requirements of several missions but only for the heaviest launch capability. This, of course, depends on the propulsion performance of the intercepting vehicle.

– Only **one** approach (nuclear sub-surface) can meet the requirements for the most unpredictable threat, a small comet impact, and only for the heaviest projected launch capability.

– All approaches greatly benefit from heavy launch capability.

It is instructive to estimate the propulsion performance that is required for a slow-push approach. The total impulse ΔI that must be provided by the vehicle is given, as function of asteroid mass and time to impact. The propulsion system will provide a jet of propellant with a characteristic velocity V_e (or specific impulse, $I_{sp} = V_e / g$, where g is the Earth gravity). The mass of propellant required to perform the maneuver and the total energy required are respectively:

$$M_p = \Delta I / V_e \quad \text{and} \quad \Delta E = \Delta I \cdot V_e \qquad (3)$$

For a given energy, it is more efficient to have the propellant at lower velocity, or equivalently, at lower specific energy or temperature; this is the same principle mentioned earlier about the respective benefits of X-ray or neutron irradiation for a standoff nuclear explosion. However, a lower propellant velocity also implies a higher total propellant mass, which must be launched into LEO. Thus, the propellant mass is a fundamental limitation. Assuming a maximum value of propellant mass of 100 metric tons (Tons) and a mission duration of 1 year, the total energy, thrust and power can be readily estimated as function of impulse requirements. The results of this trivial exercise are shown in Table 4.

Impulse	Isp	Energy	Thrust (1 yr)	Power (1 yr)	TRL
10^9 kg.m/s	1,000 s	$\geq 10^{13}$ J	32 N	≥ 320 kW	<5 yrs
10^{10} kg.m/s	10,000 s	$\geq 10^{15}$ J	320 N	≥ 32 MW	5-15 yrs
10^{11} kg.m/s	100,000 s	$\geq 10^{17}$ J	3.2 kN	≥ 3.2 GW	>15 yrs
10^{12} kg.m/s	1,000,000 s	$\geq 10^{19}$ J	32 kN	≥ 320 GW	> 25 yrs

Table 4: Slow-push propulsion performance requirements, limited by 100 Tons propellant mass.

The last column indicates an estimated technology readiness that includes the time to demonstrate a power and propulsion concept with these characteristics. The first case appears to be feasible within a relatively short time; the specific impulse is easily achieved by electric propulsion (EP) systems, and the power can be obtained by scaling-up current technology (multiple thrusters and RTGs), although nuclear fission power would be better suited. The thrust/power ratio of 100 mN/kW is somewhat larger (50%) than the current SOTA, but could potentially be achieved with a focused R&D program. An alternative approach would be a nuclear-thermal reactor (NTR), which could provide the required Isp and much more than the required power. In the second case, the power requirements are definitely above RTG-class and absolutely require a high-power nuclear reactor. The specific impulse requirement is too high for NTR, and requires an advanced EP device; a plasma thruster concept with this type of performance figures is currently in the early stages of development at AFRL. Therefore, the time-frame for development of approximately 10 yrs is feasible, provided sufficient R&D resources are available. The 3rd and 4th categories are much more difficult (even the last category is below the scenario "F" of Figure 4!); the power requirements are above nuclear fission, and while there can be thruster concepts with the required specific impulse, none of them are capable of the thrust levels that are simultaneously required. Fusion propulsion appears to be the only viable option, pending some technological surprises (i.e. new physical effects). Such concepts have been studied for a number of years, but would require massive R&D investments before actual development and demonstration. Even so, the development time would be at the very least a couple of decades, and likely to be more. It should also be pointed out that: (a) fusion reactors would need to be started by some other power source, and are likely to also require a fission reactor, or more likely to be hybrid fission-fusion designs; (b) fusion reactors very likely require significantly larger system masses in orbit than fission, and would be part of a strong space infrastructure (assembly, transport, manufacture).

Another approach to slow-push is to use the mass of the NEO itself as propellant. In that case, the mass limitation of 100 tons of propellant is no longer valid, and Table 4 can be revised. As before, the most efficient approach to deliver the momentum is to limit the propellant velocity. Thus, it would be efficient to simply "mine" the NEO as fast as possible and simply throw the mass overboard with minimal expenditure of energy[11]. In the absence of such technology, one could consider for example a radiative coupling, e.g. laser ablation of the material. Clearly something must be known about the material properties in order to achieve a good coupling (i.e. preliminary characterization mission). The other requirement concerns power. Assuming for example that the material can be heated and expanded to yield a velocity of 10 km/s (Isp=1,000 s); for $\Delta I = 10^9$ kg.m/s and a 1-yr mission time, the power delivered to the NEO is approximately 3.2 MW. Conversion efficiencies of lasers are relatively low, and even with a better than SOTA figure of 10%, this actually implies an electrical[12] power of 32 MWe – clearly a requirement for nuclear power. Larger NEOs would require excessively high powers (e.g. 32 GW for $\Delta I = 10^{12}$ kg.m/s). Note also we have not discussed the heat rejection problem at high power, another key limitation. The current SOTA figure-of-merit for power generation is about $\alpha \approx 20$ kg/kWe; this is mostly driven by radiator mass at high power. A hard-driven R&D program could potentially bring this figure down to 7 kg/kWe, and the most optimistic assumption would be of the order of 3 kg/kWe. Any power generation system will inevitably have a limited efficiency; if the system must reject 1 GW of thermal power, at the very best the radiator would weigh about 3,000 tons. Thus, extremely high power generation (>100 MWt) would require a combination of tremendously high efficiency, extremely low α, and the ability to construct, assemble and deploy large-scale structures in space.

If the latter is a technological possibility, other options may be considered. For example, a solar concentrator (Fresnel length) could be used to deliver the radiative energy

---

[11] Interestingly, this approach is exactly what is needed to drill into the core and bury a nuclear device.
[12] Chemical lasers are ruled out because of excessive reactant mass requirements.

equivalent to the afore-mentioned laser. At 1 AU distance (Earth orbit), the solar flux is approximately 1.4 kW/m^2, and to deliver the 32 MW of radiative power, the solar concentrator must be approximately 150m×150m. Unfortunately, the solar flux decays as $1/r^2$, the square of the distance to the Sun; therefore at 10 AU, the concentrator must have a linear size of 1.5 km. Nevertheless, the concept is potentially feasible, and its further development will require the ability to deploy and stabilize very large-scale structures in space, a technology that can readily be included in a systematic program for space utilization, along with high-power nuclear power and advanced propulsion systems.

III.C-3. Launch Requirements

So far it was assumed that a launch and a mission proceed successfully, and that all launch windows are available. In fact, a more realistic estimate of launch windows (Figures 7,8) shows that the probability of successful intercept can be greatly decreased even before the vehicle lifts-off. For deflection methods requiring long duration missions (slow push), the probability of failure increases with the duration, i.e. with the "gentleness" of the approach. To maintain a high probability of overall mission success, several "back-up" launches may be required, even before the first mission is completed (there may not be sufficient time to wait until the first mission fails to launch another one). This increases the overall cost of the slow-push approach, unless there is already redundancy of key assets to perform the missions. Thus, an extensive space infrastructure that includes high-performance OTVs, refilling stations, automatic/robotic assembly operations, etc would greatly contribute to reducing the risk of annihilation by NEO impact. As part of this space infrastructure, the ability to have low-cost, high-frequency heavy launch capability is another important contributor to NEO impact risk reduction.

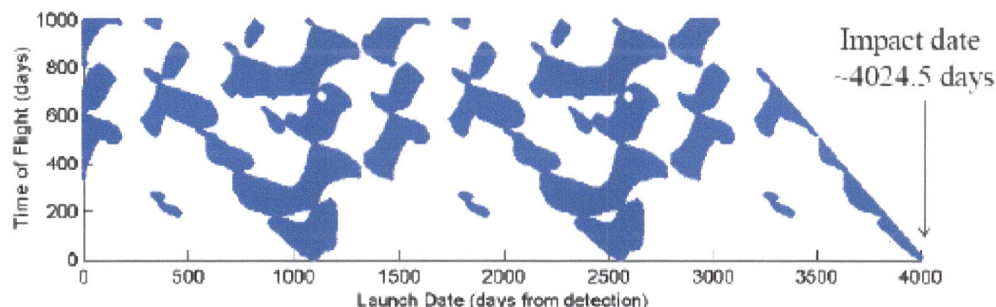

Figure 7: Launch opportunities for optimistic launch constraints (taken from [1])

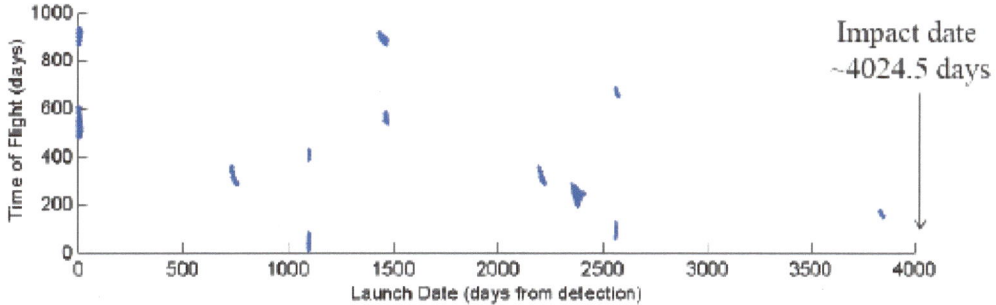

Figure 8: Launch opportunities for realistic launch constraints (taken from [1]).

17

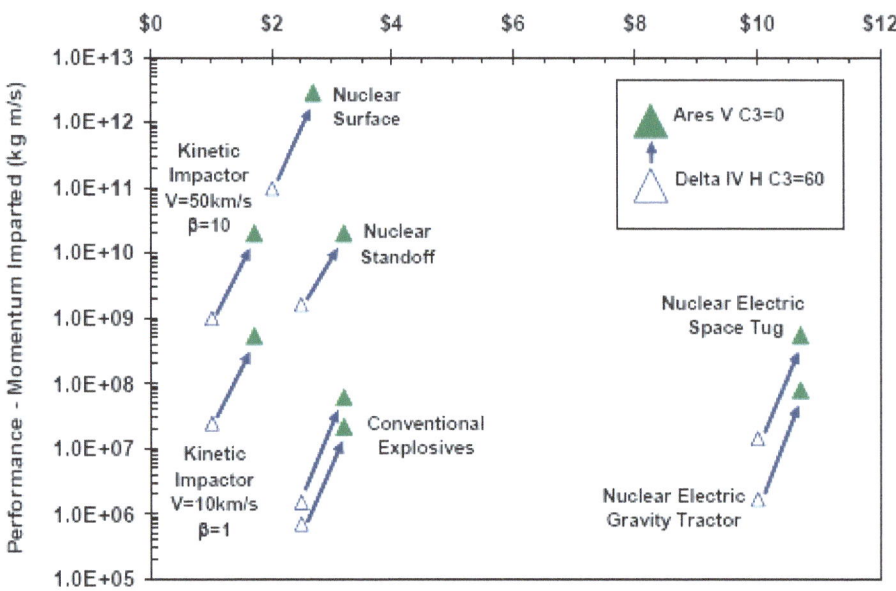

Figure 9: Deflection performance versus development and demonstration cost (taken from [1]).

Figure 9 shows the projected performance of the deflection approach versus the development and demonstration cost, in $B. It is not clear from this Figure or from the NASA report what is included in this cost estimate. Obviously, it does not include the development cost of the ARES launch vehicle itself, or the cost of developing new, high-yield nuclear explosives. A multi-year NASA program designed to demonstrate an intercept with an asteroid is also unlikely to cost less then $2B, given for example that the Mars rover mission launched in 2003, a mission with much lower technological risks, has accumulated $900M in costs since its inception. Therefore, one must be cautious in drawing conclusions about program cost from this figure.

IV. Space Infrastructure (the long-term connection)

We have alluded in the previous sections that considerable leverage could be obtained for the NEO mitigation mission if a significant "space infrastructure" exists. What do we mean by this? There are several key technologies and capabilities that can be brought to bear in NEO mitigation:

– Heavy-launch capability: this obviously facilitates the deployment of the vehicles and payloads for NEO characterization and mitigation missions, but also the deployment of space telescopes (visible and IR) and space-based radar arrays. This launch capability must be highly <u>reliable</u>, especially for mitigation. In the worst-case scenario of a comet-like impact with limited advance warning, it is critical to launch as rapidly as possible with extremely low risk of failure. The same heavy launch capability can be used for NASA missions to the moon, development of space tourism and other commercial activities, and advanced DOD missions (force projection, SBR, space-based missile defense).

– Space nuclear power: multi-MW electrical power from nuclear fission reactors will play a key role in the deployment of large platforms for Planetary Defense as well as exploration, commercial and defense missions. For example, nuclear reactors can power high-performance OTVs, provide beam power for high-altitude DOD missions, SBR and missile defense operations. Within this category one could eventually include fusion power in the far future.

– Large Structure assembly: such platforms can be used for phased-array radar, solar concentrators, and large radiators for very high power (100 MW-class) platforms. Such large structures could also play a dual role; for example, a very large array at L1 could be a phased-array radar, and very large solar power station for large-scale commercial power to be beamed to Earth, and a screen that reduces the solar flux to the Earth and reduce the effects of global warming. Such concepts are viable only if both transport (see the two previous items) and assembly can be performed reliably and at low cost. The development of robotic technology, self-assembling smart structures, redundant and self-repairing systems for long-term presence in the space environment, is an absolute requirement for this capability.

The items listed above describe the leverage of a long-term, systematic space exploration and *utilization* program which can have in facilitating Planetary Defense. Conversely, a long-term Planetary Defense program yields benefits towards a space *utilization* program:

– Asteroid mining: the same technology that may be required to drill into the core of an asteroid to plant a nuclear device would be an essential first step to mining the same object for essential elements and building blocks for space colonization. The capture and processing of the mined materials is an advanced technology that will also require full automation and large amounts of power.

– Asteroid capture: deflecting the asteroid may lead to modifying its orbit to bring it into an Earth-centered or moon-centered orbit to bring raw materials closer for use, or as the anchor mass for space elevator concepts. This may be, however, a difficult mission to perform, and one that is likely to bring trepidations, since errors in trajectory modification may precisely bring about the danger that a Planetary Defense program intends to eliminate. This mission may be more acceptable once deflection missions have been repeatedly demonstrated.

– Terra-forming: if comet-like objects or ice satellites from the rings of the gas giants could also be deflected and made to impact at precise locations (e.g. Mars), water ice could be brought to initiate terra-forming.

Although these applications may seem far-fetched to some, they are within the realm of possibilities, albeit very long-term. Yet the Planetary Defense and the new NASA "Return to the Moon" programs are essential first steps in that direction.

The "space infrastructure" is similar in some respects to the infra-structure developed by the U.S. government that facilitates commercial and national security operations, e.g. road and rail network, shipyards and harbors, airports, communication networks, etc. In space there are no roads, but a space transport and a space power network can play a similar role, using low-cost and/or re-usable access to space, long-duration OTVs, power generation/collection station and beaming, radar and optical/IR tracking stations, and refuel/repair robotic stations. This type of evolved infrastructure goes well beyond exploration missions but is truly a first step towards space utilization and exploitation of natural resources (e.g. [11]). Commercial presence in space is in its infancy and can progress only as far as the infrastructure allows it. In these early days of space utilization, national security, planetary defense and protection of commercial interests still play the most important roles. Therefore, it is logical that the DOD be a key player in the development of this infrastructure, at least in the early stages.

Within this long-term context, there are a number of key components of a space infrastructure which <u>must</u> be developed, and for which the DOD is particularly suited in taking a leading role, at least initially, due to National Security needs. These are the following

- **Component #1**: Low-cost, reliable launch. The exploration missions typically conducted by NASA are not sufficiently frequent to drive significant reductions in launch costs, and commercial activities have not yet reached a critical mass to become an economical driving force. However, DOD missions can be the dominant factor. For example, rapid reconstitution of US space assets after a surprise attack would require a high-frequency ("surge") of launches into LEO and GEO. This requires operational procedures such as rapid launcher assembly, payload matching, and automatic launch and trajectory control. If highly reliable launchers already existed, with highly modular design (multiple booster configurations easily strapped-on for variable payload/orbit requirements), robotic assembly and large-scale routine manufacturing of the launcher components, the problem of rapid reconstitution would be much easier. Clearly this goes beyond the pace and approach of NASA operations. The detailed technology does not need to be specified yet, since competing approaches may be useful, i.e. from vertical launches with no re-usable components to a fully re-usable horizontal launch vehicle. The latter could also be leveraged from technology developed for hypersonic, long-range airplanes, even up to their use as a 1st-stage. By focusing on increased reliability and reduced cost, the DOD would satisfy key requirements for Planetary Defense and greatly stimulate commercial space development (including reducing insurance costs). The issue of heavy payload capability must be addressed immediately for Planetary Defense; thus, it may be that while NASA develops the ARES-V launcher, the DOD could focus on improving the design to increase modularity, automate operations, and increase component reliability. Whether this approach, or continued and parallel development of the Delta-class of launchers, or yet another approach is chosen depends on their respective merits within the framework of a long-term plan; such comparative studies and planning shoud be done with all urgency.

- **Component #2**: Long-duration high-power OTV. These vehicles would be powered by nuclear reactors and have advanced propulsion systems capable of both high thrust and high specific impulse; they would fill the requirements of the first two rows of Table 4 shown in the previous section. As such, they would be essential components of the Planetary Defense campaign, allowing not only the "slow-push" of a large number of possible NEO threats, but also the launching of multiple characterization missions towards their targets in deep space. Such routine operations by high-performance OTVs would also have major implications for National Security, since these "space-tugs" could routinely pick-up satellites from LEO after launch (see component #1) and place them in the proper orbits, or bring back valuable assets for repair/enhancements (see component #4 below). They could also be used to push a large number of picosats for observation and monitoring of other assets, or for self-assembly into large structures (see component #5). Finally, such OTVs would greatly facilitate the current NASA mission for permanent occupation of the Moon and commercial activities in space (asteroid mining, space tourism, power generation stations). The development of this component requires nuclear space power technology, as the power requirements and the spacecraft trajectories preclude solar power. Nuclear space power has been developed through several decades, and operationally demonstrated by the former Soviet Union. A joint DOE/DOD/NASA multi-disciplinary effort can yield a new class of reactor designs with higher performance, longer operational lifetime and very high safety requirements, using the most advanced technologies available (e.g. novel materials from nano-technology).

- **Component #3**: Power generation/beaming. These platforms play multiple key roles, collecting solar power and concentrating it to ablate material from an asteroid for a slow-push, or converting it into electricity and beam it to Earth, to vehicles in transit or space settlements. The deployment of very large-scale solar power stations could then have the benefit of commercial electricity generation (beaming power to Earth), while enabling space transport and Planetary Defense, and could possibly be used as a sun-shield to reduce the impact of global warming. The nuclear reactors of the OTVs (component #2) can also serve a dual-purpose and beam the electrical power to other satellites or vehicles. Of particular interest would be very high-altitude hypersonic vehicles (recon or bombing missions) using air-breathing electric propulsion systems, powered by the microwave beam from an OTV's nuclear reactor in a high-altitude, nuclear-safe orbit. This would allow such vehicles to fly with unlimited range and loiter indefinitely, as well as having enough power for directed energy weapons, without having to place a nuclear reactor within the vehicle itself – a concept that is surely bound to raise objections. The beamed power can also be used to power that vehicle for orbit insertion, thus also playing a key role in routine, low-cost access to space (component #1). For Planetary Defense, the ability to generate highly-directional microwave beams for power transmission is immediately related to space-based radar and asteroid tracking at long distances. Thus, the same basic technology can be used for deep-space tracking and power beaming to DOD vehicles. One may also consider "relay-stations" over a deep-space network to extend the range and accuracy of the tracking. A similar network in the Earth vicinity would increase redundancy and coverage of the DOD hypersonic vehicles or launchers mentioned above. The same approach could also be used, for example, to beam power from a very large solar collector at L1 towards Earth to provide pollution-free commercial power.

- **Component #4**: Robotic/AI operations. Automatic refueling of satellites and OTVs is another key step towards the space infrastructure development, and preliminary efforts in that direction have been under-way (DARPA). With appropriate system design, robotic mechanisms and AI software, there would be no need for manned operation (i.e. no "station attendant"). Combined with the low-cost launch of supplies from Earth (component #1), the on-orbit refueling stations are an important early step towards infrastructure development. Eventually, the same procedure could be applied in reverse, i.e. receiving raw materials from asteroid or Moon mining operations and transferring them into a vehicle bound back to the Earth surface. Repairing and re-furbishing satellites and transport vehicles would be the next step; new system components (e.g. optics, solar cells, batteries, and antenna), shielding, or nuclear fuel for space reactors could be inserted at the station. Although these procedures appear complex enough to necessitate human control, it is not unconceivable that specialized robots and advanced AI could lead to completely un-manned operations. Such operations would of course have an impact on DOD missions as well as civilian or international exploration missions. The use of an international space station to perform such operations for U.S. military systems would be very problematic; thus, it would be highly advisable to develop the necessary robotic and AI technology to perform these operations in a smaller station, and in a much more cost-effective manner. The same technology can of course be applied to commercial space operations, permanent space settlements and space resource exploitation (component #5). Robotic technology is also needed to drill and bury nuclear devices in the NEO and perform assembly functions of any other concept for mitigation (laser, sail, concentrator, etc.).

– **Component #5**: Large-scale assembly/manufacturing. Some of the concepts for Planetary Defense and space utilization inevitably imply the deployment of very large structures in space. These are, for example, phased-array radars, very high-power solar collectors, highly directional arrays for power beaming and receiving/relay stations. These can be constructed from pre-manufactured modular components launched from Earth and transported to the desired location. These structures have a relatively simple pattern and can be assembled through simple rules, adequate for early phases of robotic and AI technology (component #4). Early phases of large-structure deployment, with implications for DOD missions, also include tethers, "nets" and membranes. These can be used for grappling satellites, protection against ASATs, very large optics for telescopes, space radiators, momentum-exchange boosters (using for example a small captured NEO for anchor), "bags" for raw materials, etc. Other large-scale structures, at increasing levels of complexity include space and lunar settlements ("habitats") and asteroid mining and material processing ("factories"). This is the last critical step for space colonization.

The reader may have noticed how the various components of this long-term strategy are strongly interconnected. It is more difficult to conceive the rationale for each one considered separately, but as part of an integrated scheme, the return on investment becomes much more significant. For example, there would appear to be little justification for a significant effort in developing highly reliable heavy launchers with rapid turn-around; the commercial market does not exist to support it, and the frequency of launches for DOD and NASA missions, even considering Moon permanent settlement, are not entirely sufficient. Yet if large-scale structures must be deployed for advanced DOD, Planetary Defense and future space colonization missions, the need and ROI become much more evident. Such structures may be considered impossible or much too expensive to build, despite reducing launch costs, but with advanced, nuclear-powered space tugs and advanced robotics, these are brought into the realm of the feasible. Promoting such an integrated strategy is a challenge for institutions used to funding cycles lasting only a few years; yet other countries appear very aggressive in formulating and openly intending to implement such long-term strategies. This constitutes a particular threat to the future welfare of the U.S. that should not be left unanswered.

V. Conclusions

It is sobering to realize that we have no capability at the present time to deal with the threat of a comet-like impact; should there be such an object discovered nothing could be done to avert the resulting extinction-class event – despite misconceptions popularized by the entertainment industry. For this class of impulse requirements, only buried nuclear explosives could be effective, but we currently do not have suitable launcher and space vehicle systems capable of intercepting the object. If the threat is a solid-core NEO, we do not have the capability to drill into the object. If we attempt a nuclear stand-off deflection, we do not have nuclear devices with sufficient yield…It would therefore seem very prudent to develop the required capabilities as soon as possible. The threat from smaller objects can then be handled by the same capability, while more cost-effective approaches can be designed and demonstrated throughout a longer-term campaign. Fortunately, a significant amount of technology being developed principally for the DOD and NASA can be leveraged as a starting point in such a campaign. If humanity or national survival are not a sufficient rationale, the concern over "wasting" resources for a Planetary Defense program can be greatly reduced when considering the impact of the needed concepts and technologies required for advanced DOD missions and in the long-term, for space colonization and exploitation. A coherent, long-term strategic plan must be decided with utmost urgency, for the completion of various critical phases of a Planetary Defense program. Some key first steps include the following:

- Need to (at least) stockpile current multi-megaton nuclear weapons, and preferably design and develop much higher yield devices, such as 3-stage weapons (100 Mt and above). These weapons are obviously not in stock and the designs should be optimized for radiative energy transfer (standoff deflection), until the technology to drill into various asteroid and comet cores is developed, in which case the 10 Mt-class may be sufficient. This would of course be highly controversial and raises the issue of testing. Given that the desired yield for stand-off deflection for comet-like impacts is much larger than the largest weapon tested by the U.S., it seems prudent not to rely exclusively on numerical predictions; however, such weapons could not be tested on Earth, even under-ground, and in-space testing would require a revision of international treaties.

- Need for very reliable heavy-lift launch capability, beyond heavy Delta IV capability. The ARES-V launcher being developed by NASA may provide just enough payload capability to launch a nuclear weapon of about 200 Mt (mass of 80 tons). To obtain much higher yields for a stand-off deflection of a comet-like impact (extinction-class, little warning), multiple launches would be required. The risk of failure would increase correspondingly, and the launchers may not be available in time. Thus, there is also a critical need for rapid manufacturing, assembly and stockpiling of launcher components; this requirement matches a large-scale and long-term strategy for future DOD missions and space utilization.

- Need to develop multi-MW nuclear power, with multi-year (>10) lifetime; this is critical to multiple applications, from Planetary Defense, DOD missions and space colonization. Such technology is significantly more advanced than prior efforts at deploying space nuclear power, notably for efficiency (target: > 50%) and mass efficiency (α<5 kg/kWe). Yet a consistent, long-term R&D effort can bring the technology towards these challenging figures of merits, and make nuclear space power a formidably attractive technology option.

- Need to develop MWe-class high Isp, high-thrust propulsion systems. This type of R&D is currently under-way and needs to be accelerated; however, its true potential will not be realized until the power source (item above) can be developed as well.

- Need to develop capability to deploy very large (> 1 km) structures: radar antennas, "nets", solar collectors, and beam power antennas. The latter are especially useful as a component of a space infrastructure, allowing innovative concepts for challenging DOD missions, reducing launch costs and allowing the expansion of human presence in space. These large-scale structures will need to be assembled by advanced robotic operations and artificial intelligence. In the long-term such complex, large-scale structures can be used to exploit the natural resources within the solar system.

- Need to improve the longevity of spacecrafts. This is required because of the long-duration missions for observation, tracking and deflection of the NEOs. Advanced materials, shielding concepts and embedded multi-functional structures are required, but this capability can be also augmented by redundancy and self-repairing, or by the deployment of repair/refurbishing stations.

Thus, Planetary Defense presents some unique operational and technical challenges. Yet when considering the context of a long-term space infrastructure development, the requirements for a successful Planetary Defense campaign provide multiple opportunities to bring forward the concepts and technologies that will immediately impact DOD missions across the national security spectrum, and allow future long-term commercial and societal benefits. The DOD can play an important leading role in the initial design and implementation of this long-term strategy, leveraging other key agencies such as DOE and NASA. Other leveraging opportunities are likely to be found as the strategy matures – for example, the opportunity to test AF picosat technology on asteroid characterization missions, similar to DARPA's Clementine mission[13]. It is clear that other nations have ambitious long-term goals to implement such strategies. While this may be considered as an opportunity for international collaboration, it also implies that the U.S. must take important steps forward as soon as possible if it wants to remain competitive, or even be allowed to play a major role in such international endeavors.

References:
[1] 2006 Near-Earth Object Survey and Deflection Study, Final Report (DRAFT Pre-decisional Material), National Aeronautics and Space Agency, http://www.nasa.gov/pdf/171331main_NEO_report_march07.pdf
[2] D.J. Asher, M. Bailey, V. Emel'yanenko and B. Napier, "Earth in the Cosmic Shooting Gallery", *Observatory*, **125**, 319-322 (2005).
[3] http://munnecke.com/papers/lse.htm
[4] http://en.wikipedia.org/wiki/When_Worlds_Collide_%28film%29
[5] S.P. Worden, "NEOs, Planetary Defense and Government – A View from the Pentagon", http://abob.libs.uga.edu/bobk/ccc/ce020700.html
[6] S. Glasstone and P. J. Dolan, *The Effects of Nuclear Weapons*, 3rd-ed., U.S. Dept. of Defense and U.S. Dept. of Energy, 1977.
[7] D. B. Gennery, "Deflecting Asteroids by Means of Standoff Nuclear Explosions", 2004 Planetary Defense Conf., AIAA 2004-1439.
[8] K. A. Holsapple, "On Nuking Menacing Asteroids", 34th Lunar and Planetary Science Conf., League City, Tx, March 17-21, 2003.

---

[13] This particular opportunity has been mentioned before, e.g. [12].

[9] T. J. Ahrens and A. W. Harris, "Deflection and Fragmentation of near-Earth Asteroids", *Nature* 360, 429-433 (Dec. 1992).

[10] K. A. Holsapple, "An Assessment of our Present Ability to deflect Asteroids and Comets", AIAA 2004-1413.

[11] "Space-Based Solar Power as an Opportunity for Strategic Security", Report to the Director, National Security Space Office, Oct. 2007.

[12] Col. J.M. Urias, I. M. deAngelis, Maj. D.A. Ahern, Maj. J.S. Caszatt, Maj. G.W. Fenimore III, M. J. Wadzinski, "Planetary Defense: Catastrophic Health Insurance for Planet Earth", Research Paper to *Air Force 2025*, Oct. 1996.

Acknowledgments:

The authors would like to acknowledge P. Murad and Lt. Col. P. Garretson for some stimulating discussions regarding this document.